DIVIDE and RIDE

by Stuart J. Murphy • illustrated by George Ulrich

HarperCollins*Publishers*

LEVEL 3

To Pria and Meera—
who always have enough helpful ideas to divide
between the two of them.
—S.J.M.

To Matthew, Jennie, and Peter George
—G.U.

The illustrations in this book were done in watercolor and
pen and ink on Strathmore Bristol Board.

HarperCollins®, ✎®, and MathStart™ are trademarks of HarperCollins Publishers Inc.

For more information about the MathStart series, please write to
HarperCollins Children's Books, 10 East 53rd Street, New York, NY 10022.

Bugs incorporated in the MathStart series design were painted by Jon Buller.

DIVIDE AND RIDE
Text copyright © 1997 by Stuart J. Murphy
Illustrations copyright © 1997 by George Ulrich
Printed in the U.S.A. All rights reserved.

Library of Congress Cataloging-in-Publication Data
Murphy, Stuart J., date
  Divide and ride / by Stuart J. Murphy ; illustrated by George Ulrich.
     p.   cm.   (MathStart)
  "Level 3"
  Summary: Teaches division as a group of friends goes on different carnival rides.
  ISBN 0-06-026776-3. — ISBN 0-06-026777-1 (lib. bdg.). — ISBN 0-06-446710-4 (pbk.)
  1. Division—Juvenile literature.   [1. Division.]   I. Ulrich, George, ill.   II. Title.
III. Series.
QA115.M865   1997   95-26134
513.2'14—dc20   CIP
                                                                                    AC

4   5   6   7   8   9   10
❖

DIVIDE and RIDE

It's Carnival Day today
for our group of 11 best friends.

The Dare-Devil's first, and we'll need to divide.
2 people fit in each seat,
and each seat must be filled before we can ride.

We can fill 5 seats,
but 1 friend is left over
from our group of 11 best friends.

11 divided by 2 = 5 full seats

. . .with 1 friend left over.

Amanda yells, "Come fill this seat!"
to a kid we don't even know.

$11 + 1 = 12$

12 divided by 2 = 6 full seats.

9

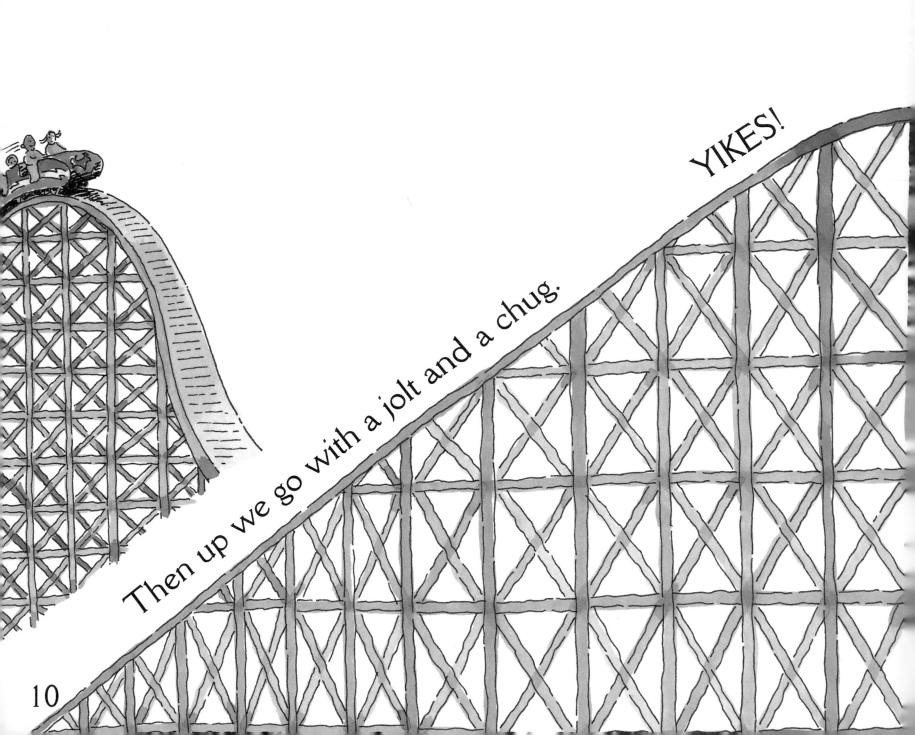

Then up we go with a jolt and a chug.

YIKES!

10

We're flying straight down!

For the Satellite Wheel,
again we'll divide.

It holds 3 people per chair,
and each chair must have 3
before we can ride.

We can fill 3 chairs,
but 2 friends are left over
from our group of 11 best friends.

11 divided by 3 = 3 full chairs

. . .with 2 friends left over.

13

Patti and Jack shout, "Jump in here!"
to a kid we've never even met.

14

$11 + 1 = 12$

12 divided by 3 = 4 full chairs.

Then up we start. We swing to the top.

16

Then 'round and around we go!

17

For the Twin-Spin Cars, once more we'll divide.
The ticket man shouts, "4 people per car.
Each car must be filled before you can ride."

We can fill 2 cars,
but 3 friends are left over
from our group of 11 best friends.

4 people per car!

11 divided by 4 = 2 full cars

. . . with 3 friends left over.

Mickie, Jill, and Rob scream, "Come with us!"
to a kid we've never even seen.

11 + 1 = 12

12 divided by 4 = 3 full cars.

Then we turn and twirl. We twist and whirl,

22

and we spin around real fast!

23

At last! On the raft, we don't need to divide.
There are 14 seats in all,
and every seat must be filled before we can ride.

To fill all the seats
we need to add 3 kids
to our group of 11 best friends.

$14 - 11 = 3$ empty seats.

So we call to the kid
we didn't even know,

we yell to the kid
we'd never even met,

and we holler to the kid
we'd never even seen before.

27

Then we start to slide, and to slip and slosh.

We land at the bottom

28

with a great big SPLASH!

Now we're 14 carnival friends!

11 + 3 = 14.

f you would like to have more fun with the math concepts presented in *Divide and Ride*, here are a few suggestions:

- Read the story together and ask the child to describe what is going on in each picture. Ask questions throughout the story, such as "Which ride would you most like to go on? Why?" and "If eleven friends sit three people per car, how many friends are left over?"

- Encourage the child to tell the story using the math vocabulary: number of kids "per" seat, "divide," and "left over." Introduce words such as "groups of," "sets of," and "remainder."

- Talk about familiar settings in which large groups are divided into smaller groups, such as teams for a game, rows of seats on a bus, and people at each table. Make sketches or diagrams of each situation. Discuss what happens with those who are left over. Is a new team formed? Do some people have to stand on the bus? Is an additional chair added to the table?

- Draw the 11 best friends using stars as shown on the math summary pages, or use pennies or pebbles to represent the friends. Together, practice grouping the friends into sets of 2s, 3s, and 4s. Are there any friends left over? How many?

- Take another look at the story. What if the group of 14 new best friends went to the carnival together? How many seats would they fill on each ride? Would there be any friends left over?

Following are some activities that will help you extend the concepts presented in *Divide and Ride* into a child's everyday life.

Having a Snack: Invite a group of friends for a snack. Take out a box of cookies or some raisins or nuts. Ask the friends to divide the food evenly. How many cookies, raisins, or nuts does each friend get? Are there any left over?

Shopping: Set up a play candy store and give the child a handful of pennies. Price some candies at 2 cents, 3 cents, 4 cents, and 5 cents. How many 5-cent candies can be bought with the pennies? How many 3-cent candies? Are there any pennies left over?

Card Games: Play a game of cards with some friends. Deal out the cards evenly. How many cards docs each person get? Are there any left over?

The following books include some of the same concepts that are presented in *Divide and Ride*:

- THE DOORBELL RANG by Pat Hutchins

- A REMAINDER OF ONE by Elinor J. Pinczes

- ONE HUNDRED HUNGRY ANTS by Elinor J. Pinczes